Like people, most animals ha[ve] [periods]
of rest each day called sleep[. How]
each animal sleeps—when, w[here,]
and how long—depends on its habitat
and way of life.

Lions, like many
other animals,
often yawn right
before a long rest
or period of sleep.

Animals sleep in many different positions and places, for different amounts of time. Some animals sleep whenever they are not eating or grooming themselves. Others may not sleep at all or only for a few hours. Many may rest instead. There is much about the sleeping habits of wild animals that is still a mystery.

This two-toed sloth, which lives in tropical forests in Central and South America, sleeps hanging upside down from a tree branch. The two-toed sloth is active for about 8 hours a day and then spends the rest of its time snoozing.

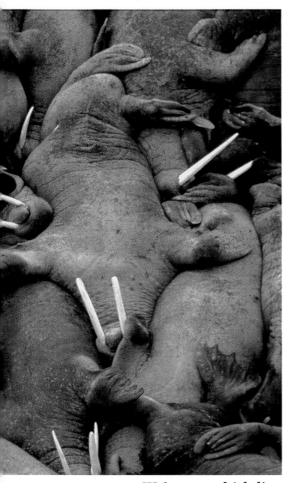

Flamingos inhabit wetlands and sleep standing on one leg with their beaks tucked behind one wing. Flamingos do most things in groups—even sleeping.

Walruses, which live only in the Arctic Ocean and nearby ice-covered areas, sleep restlessly in herds, huddled together for warmth.

Scientists don't know a lot about the sleeping habits of snakes, but they do know that they have periods of inactivity. A python, like the one shown here, can swallow a gazelle whole and then be inactive for up to six months digesting it.

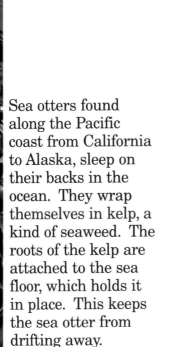

Sea otters found along the Pacific coast from California to Alaska, sleep on their backs in the ocean. They wrap themselves in kelp, a kind of seaweed. The roots of the kelp are attached to the sea floor, which holds it in place. This keeps the sea otter from drifting away.

Scientists are not sure if insects sleep, but they do know that certain insects, like bumblebees, have favorite resting sites, such as on the petals of flowers.

Predators

Predators are among the best sleepers of all animals because they worry the least about being attacked. Since large, meat-eating cats are skillful hunters, they have more time to sleep than most animals. Like many wild cats, leopards sleep in trees, or on the ground, for about 14 hours a day.

Lions sleep even more than leopards do—up to 20 hours a day! The lion shown here is sound asleep after a large meal.

Most American black bears and grizzly bears **hibernate** in sheltered areas, or dens, from 4 to 8 months out of the year, depending on how long the winter is. When not hibernating, these bears are mostly active at night and early in the morning. They spend the rest of the time resting or sleeping.

During hibernation, the bear's body temperature drops and its breathing slows down. But during warm days the bear may interrupt its sleep by taking short trips outside of the den.

The American black bear lives mostly in forested areas. Sometimes it sleeps in trees, at other times it sleeps on the ground.

Grizzlies live in forests and along coastlines. These two grizzlies in Alaska are sleeping— one on a rock in the water, ready to catch a salmon when it wakes up; the other one sprawled out in the snow on its back.

Polar bears, who prey mostly on seals and fish, live in the arctic regions around the North Pole. Sleeping a lot helps them to warm up or cool down their body temperature and to save energy between meals. Polar bears sleep in a variety of positions.

When hot, polar bears will press their bellies to the ice or sprawl on their backs with their paws up as in the big picture above.

When they are cold they often flop face first into the snow.

When a polar bear is trying to
keep warm during a cold winter
snooze, it will curl up and cover
its nose with a warm paw.

Some animals, such as gorillas, sleep peacefully because they are too large to have enemies.

Gorillas live in the lowland or mountainous tropical rain forests in Africa. Gorilla families take lots of naps, especially between feedings and at noontime, the hottest part of the day. They collect grasses and leaves to sleep on during the day, or sometimes they take a noonday snooze on a fallen tree limb.

Before it gets dark, the gorilla will build a new bed to keep itself off the moist ground. The gorilla reaches out around itself and pulls down the stalks of large-leafed plants. By joining the branches in the middle, gorillas make large bowl-shaped beds.

The rhinoceros,
the hippopotamus,
and the elephant
of Africa also sleep
peacefully.

Rhinoceroses sleep at night and rest on and off during the day, in between feedings and at noontime. Sometimes they rest in mud or a pool of water. At other times they may rest on land. Tickbirds perch on rhinos, and often serve as a warning signal or alarm if danger is near.

Hippopotamuses feed on the African grasslands at night for 5 to 6 hours. The rest of the time they wallow in water, digest their food, rest, and sleep. Hippos often use each other for pillows, as they sleep standing in the water.

Elephants can sleep either standing up or lying down. They sleep at noon and after midnight for a daily total of 4 hours. Full-grown elephants don't sleep long lying down, however, because their tremendous weight would injure their internal organs. Mother elephants often sleep standing up with their babies on the ground sleeping next to them.

Animals that are commonly hunted by predators, such as giraffes, must look out for danger, so they are usually light sleepers. They must be ready to flee at the first sign of danger.

When a giraffe rests or dozes, it stands or lies down with its head erect, ears twitching, eyes half closed. It usually rests at noon when the African sun is at its hottest. It sleeps deeply for only 3 to 4 minutes at a time with its head arched back resting on its hind legs. Adult giraffes sleep about a total of 20 minutes a day.

Zebras, deer, and kangaroos also sleep lightly because they are constantly watching for danger. Smaller animals of prey, such as red foxes and raccoons, often sleep in nests, burrows, or dens underground.

Raccoons often sleep in dens in the ground or in nests in tree hollows. In parts of the United States and Canada, they will remain in the same den and become very inactive during the winter.

During the heat of the day in Australia, the kangaroo will rest in the shade of rocks or trees. Sometimes it will dig away topsoil and flop down in the cooler soil underneath and sleep on its back.

The red fox rests in a sheltered spot, but it usually has a main den with a couple of emergency burrows connected by tunnels, where it also rests during the day.

Even when bedded down, whitetails doze fitfully. Deer are always on guard against predators. This fawn watches ahead, while her ears are tuned in to anything taking place behind.

When dozing, zebras stand with their heads drooping, their eyes half-closed, their ears pointing out to the side, and their tails swishing back and forth chasing flies. During a deep sleep they sit down with their legs tucked underneath for a quick getaway should they be awakened suddenly.

Animals of the Sea

Animals that live in the sea sleep
in many different ways. The
California sea lion has a variety
of sleeping positions—some in the
ocean, some on land.

Sea lions often sleep
propped up on their
front flippers with
their heads thrown
back and their noses
pointed straight up
in the air.

Sometimes sea lions will wriggle
up onto rocks warmed by the sun
and sleep on their bellies or on
their sides.

They can also sleep floating in the
ocean with their heads pointed up
and one flipper sticking straight
up out of the water. The flipper
absorbs heat from the sun and
helps to keep the sea lion warm.

Other animals of the sea—dolphins, parrotfish, and manatees—have interesting ways of sleeping.

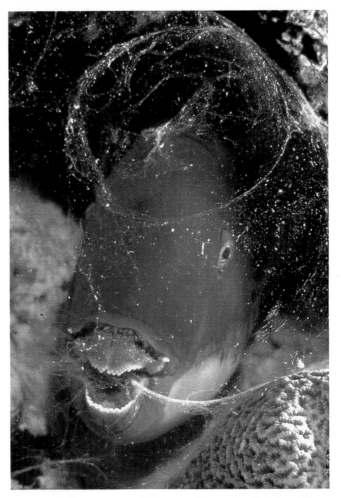

Dolphins rest in the water by shutting down one-half of their brain at a time. The other half controls movement, breathing, and keeps one eye open for predators. In total, each half rests for about 3 or 4 hours out of 24.

Parrotfish surround themselves at night with a mucus "sleeping bag" that prevents their enemies from smelling their scent. At daybreak, they will eat their sleeping bags and swim away.

Manatees rest about 6 to 10 hours a day. When they rest, they float near the surface of the water or lie on the ocean bottom. Their eyes are closed and their bodies are motionless. Manatees must surface to breathe every 2 to 3 minutes.

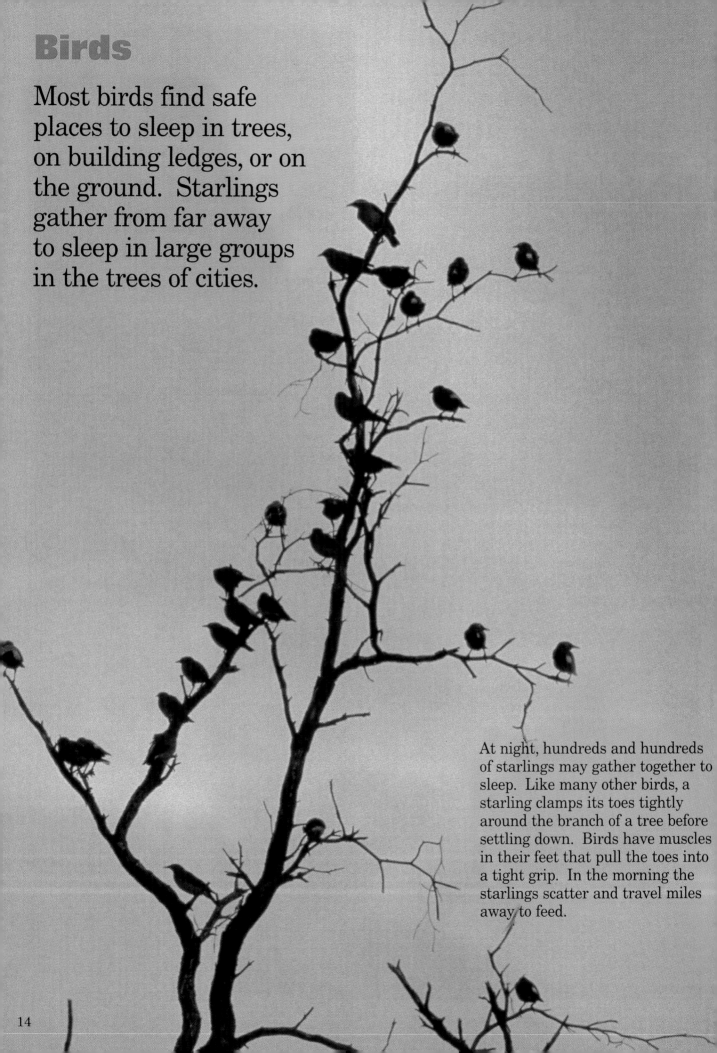

Birds

Most birds find safe places to sleep in trees, on building ledges, or on the ground. Starlings gather from far away to sleep in large groups in the trees of cities.

At night, hundreds and hundreds of starlings may gather together to sleep. Like many other birds, a starling clamps its toes tightly around the branch of a tree before settling down. Birds have muscles in their feet that pull the toes into a tight grip. In the morning the starlings scatter and travel miles away to feed.

The ostrich, owl, and quail are other members of the bird family that have interesting ways of sleeping.

Ostriches camouflage themselves from predators while sleeping by stretching their necks along the ground and folding their legs beneath their bodies, leaving only round humps sticking up. From a distance, the hump looks like a bush that grows in their habitat, the African savannah.

Owls are nighttime hunters. When day comes, they fly to their nest, usually in a dead tree trunk, and close their big eyes. Here they sleep until twilight.

Quail sleep on the ground where they face more danger than birds do in trees. To protect themselves, quail form a circle, tails together in the center, heads facing out. If an enemy approaches, each bird flies off in a different direction. This way most of them escape.

We know that animals sleep in many places and unusual positions for greatly different amounts of time. Although the sleeping habits of most animals are not like ours, many animals wake with a big yawn and stretch—just like us!